SCIENCE WORDS FOR LITTLE PEOPLE

Helen Mortimer & Cristina Trapanese

Space

OXFORD
UNIVERSITY PRESS

Observatory

Welcome to our sleepover observatory on planet Earth! It is where we watch the night sky.

Telescope

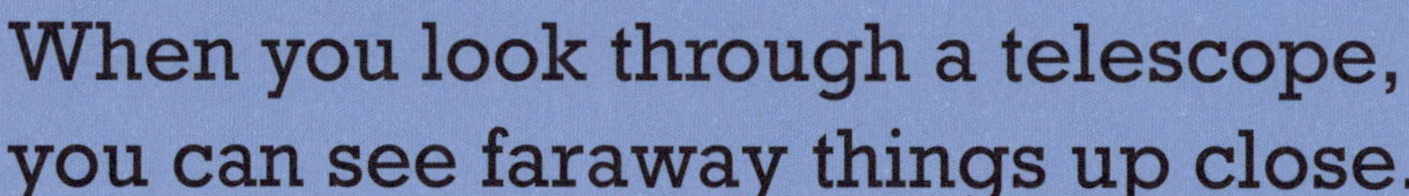

When you look through a telescope,
you can see faraway things up close.

Wow – look at these craters!

Everything in space is very, very far away!

BZZZ! Craters are dips that look like giant bowls.

Astronaut

When astronauts go into space they float around because there is no gravity to pull them down.

BZZZ! Did you know that 'astronaut' means 'star sailor'?

Solar system

There are eight planets in our solar system. They all go around the Sun in an orbit.

Saturn
Venus
Neptune
Sun
Jupiter
Earth
Uranus
Mercury
Mars

Day and night

As the Earth spins in space, it's daytime on the side facing the Sun and night-time on the side facing away.

Goodnight

Phases of the Moon

The Moon changes every night because different parts of it are lit up by the Sun as it moves around the Earth.

Half moon!
Crescent moon!

Space mission

Rockets and spaceships take astronauts into space. Every mission starts with a countdown. Mission control, are you ready to count?

We have lift off!

Stars

Stargazing is fun to do!

Twinkly stars make shapes in the sky called constellations. They all have different names.

Space probes

Scientists send special robot spacecraft called probes into space to find out things about the galaxy and the universe.

BZZZ! Our galaxy is called the Milky Way.

Robots can stay in space for much longer than people.

Aliens

Some people think that one day we may discover aliens living on other planets.

It's fun to imagine what our alien friends might look and sound like.

Blip blip!
Meep meep!

Space rocks

Chunks of rock called comets, asteroids and meteors are always whizzing through space.

Stargazers sometimes see these rocks at
night, like fireworks in the dark sky.
Look, a shooting star!
Let's make a wish!

All about space

What would you wish upon a star?
If you met an alien, what would you do?
What do you think it's like on the Moon?

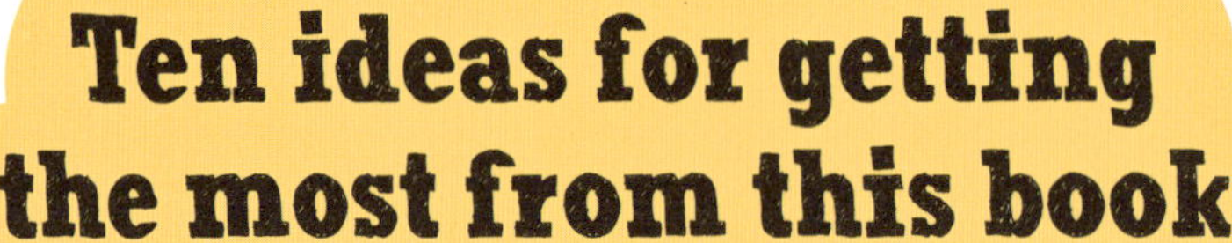

Ten ideas for getting the most from this book

1. Find somewhere cosy and comfortable to share this book. It's important not to feel rushed.
2. Remember that science isn't something that happens only in a classroom – there are opportunities all around us to talk about science concepts and use scientific language.

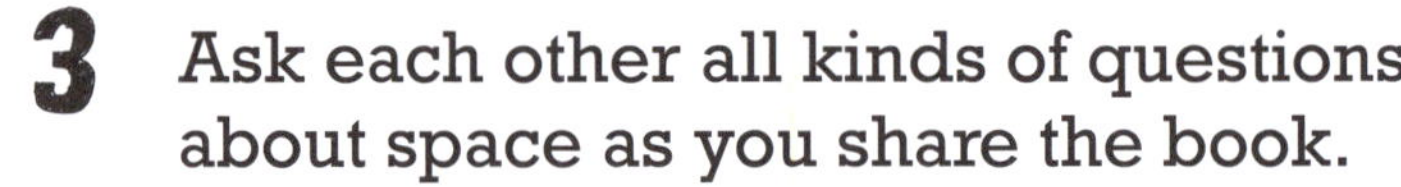

3. Ask each other all kinds of questions about space as you share the book.
4. The children in this book are enjoying a sleepover. Why not organize a space-themed play date or sleepover and try some of the space-inspired craft activities?
5. Look for space-themed picture books in your local library to share together.

6 Try looking at the night sky together to see the stars and see what shape the Moon is.

7 Make up some alien words: how would your alien say 'hello', 'welcome' or 'goodbye'?

8 Try to remember the eight planets in our solar system and make a collage picture of them.

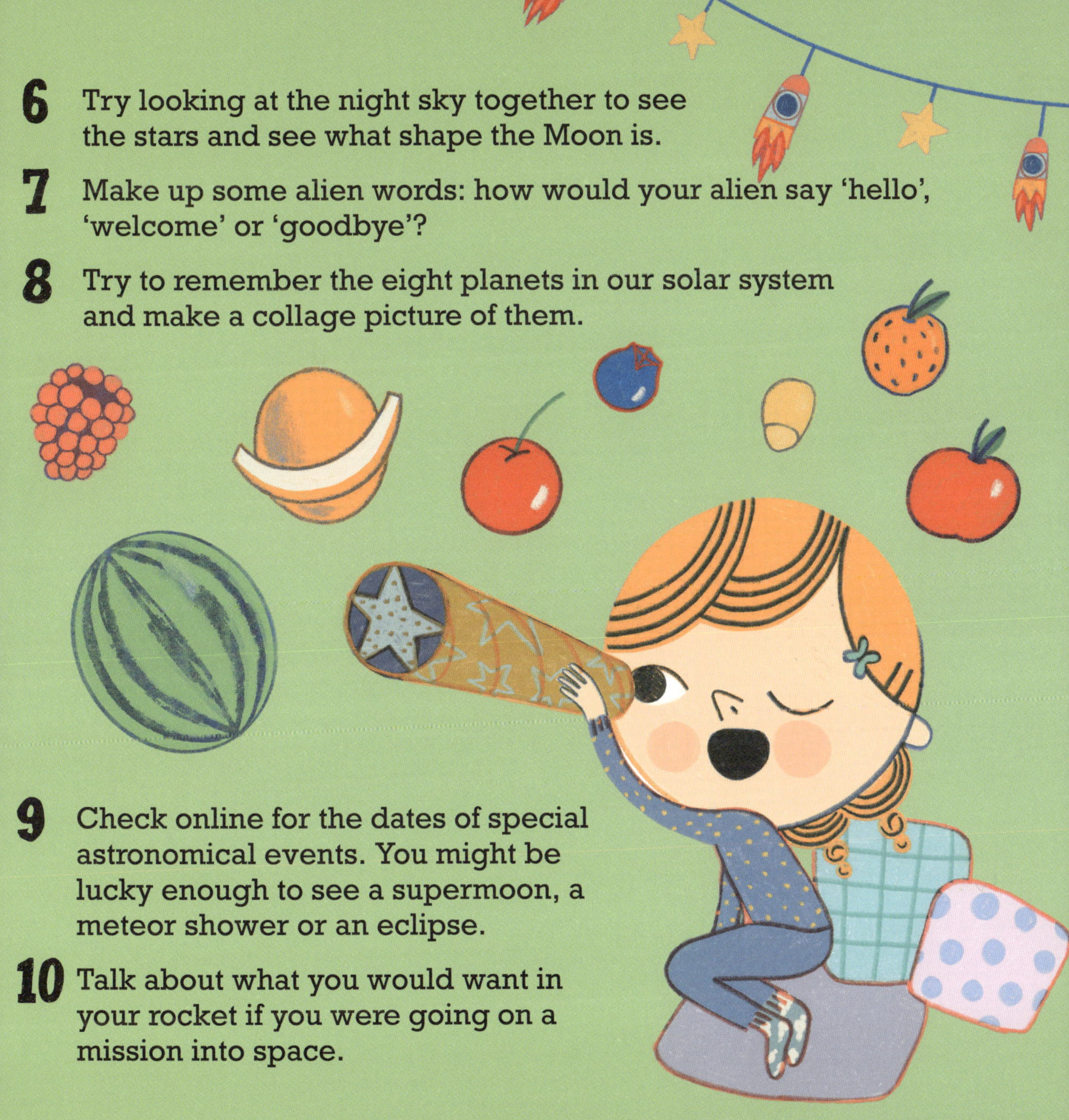

9 Check online for the dates of special astronomical events. You might be lucky enough to see a supermoon, a meteor shower or an eclipse.

10 Talk about what you would want in your rocket if you were going on a mission into space.

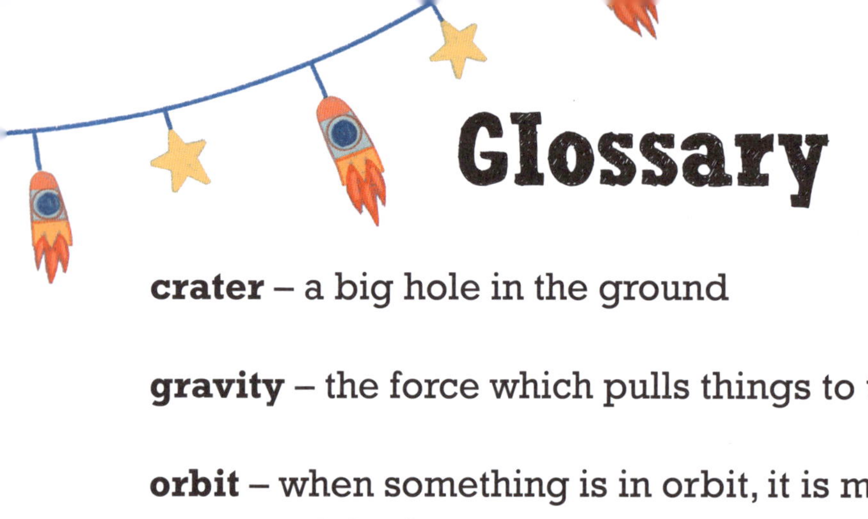

Glossary

crater – a big hole in the ground

gravity – the force which pulls things to the Earth

orbit – when something is in orbit, it is moving round the Sun or a planet

observatory – a building where you can use telescopes to look at the stars

solar system – the Sun and the planets which go around it